# ALTERNATE MEMORIES

## The Mandela Effect

## FINAL EDITION

**Jay Wheeler**

Alternate Memories – The Mandela Effect
Copyright, Alternate Memories
Copyright Content, 2021
Copyright Jay Wheeler, 2021

Authored by Jay Wheeler
Cover by Jay Wheeler
Final Edition 2021

# **CHAPTERS**

I: The Mandela Effect — Page 1

II: Examples — Page 5

III: Theories — Page 15

IV: Misdirection & Illusion — Page 27

V: Accepting the Impossible — Page 39

VI: Alternate Memories Explained — Page 43

VII: Final Thoughts — Page 69

# Chapter I:

# What is the Mandela Effect?

Many people have heard of the Mandela Effect or even seen examples. Some have even been baffled themselves by having alternate memories of actual events, but let's start at the beginning.

It began in 2013 when Nelson Mandela passed away.

In 1962, Nelson Rolihlahla Mandela was arrested and sentenced to life in an African prison. After serving 27 years, Mandela was released in 1990 by President F. W. de Klerk.

This stuck a lot of people as a surprised because many remember his untimely death in prison in the late 1980's-early 1990's. Millions of people distinctly remember news anchors covering the story, a publicized funeral, and even rioting in the streets shortly after the news of his death broke.

Despite more than half the population convinced Nelson Mandela died in prison around 1990, he was actually released from prison on February 11, 1990.

Following his release, Mandela became president of South Africa. He remained as president of South Africa for 5 years, from May 10, 1994 – June 16, 1999.

Over a decade later on December 5, 2013, he passed away from a respiratory infection. This was the Mandela Effect reached a global scale.

In 2010, Fiona Broome coined the now-famous phrase 'The Mandela Effect' and kept an online journal of examples which you can see at 'MandelaEffect.com'.

The Mandela Effect didn't gained worldwide notoriety until 2013 when Nelson Mandela really passed away. After his death was announced by the media, people were shocked because they vividly remember the events from the early 1990's and internet searches exploded, earning the Mandela Effect its popularity.

Since 2013, there have been hoards of people reporting and sharing memories of events and other things they remember flawlessly, but all these memories seem to be incorrect or inaccurate. Many of which we will explore in this book.

With the internet and other advanced forms of communication, information can be shared much faster and wider than in 1990 and as soon as a new Mandela Effect example or alternate memory is shared online, there seems to be many people having the same false memory and those who remember it accurately.

Some of the alternate memory examples reported even have a near 100% agreeance rate, meaning some of the alternate memories reported are remembered by everyone incorrectly and no one agrees that the current truth is how it's always been, forcing people to search every quadrant of their brain for a solution, but being left with nothing but frustration and mystery.

This entire phenomenon would be easily explained if it was just a few people around the world having alternate memories, but when there are literally millions of people misremembering huge events that never happened, it merits more investigation and that's what I've spent years researching to conclude and explain the Mandela Effect in this book.

There have actually been reports of people having alternate memories since the 1960's, though they weren't widely reported. There was no internet and the media outlets that did hear of them likely just swept them under the rug as they didn't feel it was a solid story.

Returning to the initial event that spawned this phenomenon, there's a book titled ***English Alive, 1990: writings from High Schools in Southern Africa***, in which you will find the following:

*"The chaos that erupted in the ranks of the ANC when Nelson Mandela died on the 23rd of July, 1991 bought the January 29th, 1991 Inkatha-ANC peace accord to nothing."*

The passage is also in ***Western Cape Branch of the South African Council for English Education***, which was published on October 1, 1991.

If you were under the impression Nelson Mandela died in prison, you're not alone. However, not everyone has this memory. Some people never heard of him dying in the late 80's or early 90's. But, don't worry, there are plenty of other alternate memories and I've yet to meet a person that has accurate memories of every example.

Some of the examples we're going to look at are memories we have all grown up with and believed. Some are popular sayings the world knows, others will probably make you question the accuracy of your own memory or think of something even more bizarre behind the cause. We'll look at the mind blowing theories in a later chapter. The Mandela Effect is just the beginning...

# CHAPTER II:

# EXAMPLES

Since 2013 when Mandela died, there alternate memories have been pouring in. The internet has snowballed the phrase into one of present day's most baffling mysteries, but I'm here to conclude that mystery and try to provide some logic to why we are experiencing such puzzling false memories.

Many of the alternate memories reported have no trace of ever being factual or having existed. Many of us have gone back to re-watch movies, re-listen to songs and revisit signs and logos that we all remember completely differently to how they are. I will admit, I was completely speechless when I discovered some of the alternate memories people shared, but after extensive research, logical thinking and my research background in neural science, the answers are much simpler than we may think.

For now, let's look at some of the top examples that have baffled the world and then we will go into explaining each and every one of them in hopes I help a few people sleep better knowing they're not in an alternate reality or parallel universe.

## *Empire Strikes Back*

Luke Skywalker's last encounter with Darth Vader in *Empire Strikes Back* was one of the most epic and memorable scenes to ever bless our screens. Not only was it one of the greatest reveals in cinematic history at the time, but it was a staple in film history as it was the end of possibly the greatest sequel ever made.

Do you remember when Darth Vader breaks the news that he's in fact Luke's father? How do you remember that line going?

Almost everyone remembers it as Luke, I am your father', but that was never the line. I can even remember my father calling me "Luke" as a joke, just so he could deliver the full line to me. Many other people have the same memory of the line, including a person we saw sharing a story of their friend naming their son "Luke" who was inspired by that specific line in the movie.

Well, this line never existed in the movie. Vader's actual line was "No, I am your father".

Many people who are baffled by this alternate memory recall an interview with James Earl Jones (the actor who voice Darth Vader) where he recited the link *"Luke, I am your father"*, which added more fuel to the fire when even the actor who played the role recalled the alternate memory and not the exact line from the movie.

It doesn't seem to matter how confident we are that the line was *"Luke, I am your father"*, if you watch any version of the movie, whether it be VHS, DVD, Laser Disc, Blu-ray, the remastered, the original releases or even if you read the script, there is not a single trace of this memory so many of us share.

## *Chick-fil-A*

Didn't the Chick-fil-A fast-food company used to be spelled "Chic-fil-A" without the "k" in the name? People remember the famous chicken franchise to be "Chic-fil-A". Many have discussed how they even joked about the name being 'Chic' (pronounced "sheek").

There've been literally millions of people who agree there was never a 'k' in the name or company logo. When first researching the Mandela Effect, this is one that caught my attention, as well as many of my friends' and we were all convinced beyond any reasonable doubt it was previously spelled "Chic-Fil-A", so I directly contacted Chick-Fil-A's corporate office and asked them when they added the "k" into their name and changed it from "Chic-Fil-A" to "Chick-Fil-A". With such great customer service in all areas they replied very quickly, but it wasn't the answer I initially expected. They told me: *"our name has always had the "k" in it."*

Despite them owning both "Chick-fil-A.com" and "Chic-fil-A.com" (the latter re-directing to their official website with the "k"), there's not a single picture, review, report or article anywhere on the internet that shows the alternate memory millions of us share. The only place you'll find the Chick-fil-A logo without the "k" is when you're searching for Mandela Effect related content.

## *The Berenstain Bears*

The book series known by millions and remembered as "The Berenstein Bears" (pronounced "Berensteen") is another popular example of the Mandela Effect which has baffled nearly everyone who grew up with the famous children's book series.

The Berenstain Bears is considered by millions to have changed from its original "Berenstein Bears" due to the mystery of the Mandela Effect. This is a good example of the high agreeance rate as there seems to be no one who remembers the book series as the name it's now called, "Berenstain", with an "a" instead of an "e". Is your mind blown yet? Don't overdo it. We're just getting warmed up.

## *Snow White and the Seven Dwarfs*

Here's another one that really puzzled me, as well as the ret of the world apparently. What do you remember the evil queen having said when she spoke into her mirror in the 1937 Walt Disney classic, *Snow White and the Seven Dwarfs*?

Wasn't it "Mirror, mirror, on the wall"? That's

what I thought, as did millions of others, but again that's not correct. Nearly everyone who has ever shared their thoughts or commented on this example remembers the line being "Mirror, mirror, on the wall", but if you rewatch the movie or that specific scene, in any format, the line is *"Magic mirror, on the wall"*.

The line "Mirror, mirror, on the wall" never existed in the animated Snow White and the Seven Dwarfs.

## *Jaws and a Bigger Boat*

In Stephen Spielberg's 1975 masterpiece, Jaws, there's a line that is said whilst they're on the boat. Most movie fans and Mandela Effect followers argue the line used to be "we're gonna need a bigger boat".

The truth is, the line in the movie is "you're going to need a bigger boat. Again, this is only a single word that's causing an alternate memory for many people, but it's earned its place on the list of the most baffling alternate memories.

Although there seems to be millions of people who are adamant the line has changed, I will explain in detail the truth about this alternate memory later in the book.

## *Interview with the Vampire*

The famous blockbuster movie, *Interview with the Vampire*, has baffled many people as they believe it was always called "Interview with a vampire". This has become one of the more popular Mandela Effect examples as so many people have this alternate memory of the movie title having an "a" instead of "the". Many vampire fans and avid readers swear that they're remembering the title "Interview with a vampire" as the correct spelling, but unfortunately they're only met with mystery and a memory that is incorrect.


## *Sex and the City*

Didn't the popular TV sitcom *Sex and the City* used to be called "Sex in the City"? Many people remember it being just that. However, if you check any sources including the show, logos, and artwork for this alternate memory, you'll not find any reference to the show being named "Sex in the City" as the "in" no longer lives in the title. It's actually spelled "Sex and the City".


## *The Monopoly Man*

We've all played the ever popular board game, Monopoly. With so many people and such time spent playing it you'd think we'd remember how the main character appears. Well, apparently none of us remember him correctly. If you're one of the few who think his appearance hasn't changed, maybe you're right.

The famous Monopoly man, known as Rich Uncle Pennybags, is previously thought to have a monocle (glass eye lens), at least we thought so. It was considered by many sharing this alternate memory as his "signature" fashion piece. If you agree and think he used to have this accessory, maybe you should go back and look at a Monopoly board or have a browse on Google because Uncle Pennybags has never had a monocle in any version of Monopoly, artwork or branding.

### *Dolly & Jaws*

In the 1979 classic James Bond movie *Moonraker* you may remember the strange love affair between Jaws (Richard Kier) and Dolly (Blanche Ravalec), but what you might not remember correctly is her appearance. Do you remember Dolly having obvious silver braces on her teeth? Almost an imitation of Jaws' signature metal teeth? Not anymore. If you watch the movie now, you may be surprised to learn that the character of Dolly never had braces in Moonraker. There were even descriptions written from years ago mentioning and describing her characters as having said braces.

## *Three Little Pigs*

Arguably one of the most well-known children's nursery rhymes of all time, Three Little Pigs is something millions of us grew up with. You may remember the rhyme going: *"I'll huff, and I'll puff, and I'll blow your house down!"*

That's how I, and many others remember it. Though to our surprise, if you read any version or recording of the classic rhyme, you will notice the line now being: *"I'll huff, and I'll puff, and I'll blow your house in!"*

Very few agree they were brought up with the line being "Blow your house in"

## *Hello, Clarice*

One of the best screen plays of all time grew into one of the greatest thriller classics in *The Silence of the Lambs*. Horror fans are rather attentive to detail as am I, so when I discovered this alternate memory, I was amazed.

If you remember Hannibal Lecter saying "Hello, Clarice" either on one of their meetings in the prison or on the phone call at the end of the movie, you're experiencing a false memory because it never happened.

At no point in the entire movie does Hannibal say "Hello, Clarice".

## *What if I told you everything you believed was a lie?*

In 1999, the world was bless with the mind-bending movie that blended artificial intelligence, simulated reality and action all into one. That movie was *The Matrix*.

After the movie's release, many recited a line that went something like... "What if I told you everything you believed was a lie". This line was allegedly said when Morpheus offered Neo the choice between the blue and red pill. Yes, you guessed it. That line was never said in the movie, nor during the pill scene or any other scene for that matter. Nor was it ever in the script or a cut scene.

### *C-3PO*

Another Star Wars memory people have is one of the greatest robot sidekicks of all time, C-3PO. No matter how many times Star Wars-crazy fans watch the movie, nearly everyone is surprised when you realize 3PO actually has a silver leg now.

### *It's the Flintstones!*

Well despite many people agreeing the famous animated series was titled "The Flinstones", it actually contains a second "t" and the title is The Flintstones.

### *Looney Toons*

With many of us either growing up with the famous cartoon series or at least seeing it and remembering the logo at the start, you probably still read the title of this example incorrectly. The show is actually titled *Looney Tunes* and not "Looney Toons".

Though I have only given a few examples, I will examine and explain many more in the concluding chapter.

With a good collection of examples that I'm sure no one managed to read without getting stung with one alternate memory, let's look at some of the theories that people have shared to try and explain what the Mandela Effect could be, what the cause could be and why it may be happening.

# CHAPTER III:

# THEORIES

How can this worldwide phenomenon be occurring? How can so many millions of people experience the same exact memories that never occurred? There's many theories, but none of which have any concrete evidence to prove they're correct.

Most just lead to more questions we're not able to answer and some of the theories lead to questions we can't even comprehend. However, it's worth mentioning a few of the main theories as they seem to have potentially possible explanations, even if they are reaching for science the greatest minds on earth can't even begin to answer.

Though some theories are wild, out there, crazy, unimaginable, they are theoretically plausible in the name of science.

There's many other theories, but I've only included a few chosen ones that I feel readers will have interest in and that are also feasible, even if we have to stretch our thinking far beyond our understanding of physics and science.

## *CERN*

The first theory we'll look at it as ambitious as the movie Interstellar, maybe even more so.

CERN is a European organization for nuclear research which created the LHC (Large Hadron Collider) which is now the most powerful particle accelerator in the world.

It has also been called the "Big Bang Machine", a 17-mile long underground ring on the border of France and Switzerland.

Its primary use is to crash subatomic particles together at almost the speed of light (186,000 miles per second), allowing scientists to sift through the aftermath in search of answers to some of the most complex questions in physics relating to dark matter, super symmetry, and higher dimensions.

Scientists also hope CERN's technology will help bridge the gap between quantum physics and general relatively, Albert Einstein's century long theory.

The LHC took decades of work to finally come together and in 2008 it did, just 5 years before Nelson Mandela passed away. Although, 9 days later it failed due to an electrical problem. Another 12 months and almost $140 million later, it was back to work.

In 2012, CERN discovered the Higgs boson, maybe known more widely as "The God Particle". A theoretical particle that was first mentioned in 1964 by Peter Higgs, one of six physicists who proposed the mechanism that suggested the existence of such a particle.

Higgs boson is the particle that gives other particles their mass. It's very important in the world of physics and it possesses seemingly magical properties where it can create mass.

The late, great, British physicist, Stephen Hawking, warned that the Higgs boson could one day be responsible for the total destruction of our known universe whereas a quantum fluctuation creates a vacuum that expands through space and time. In layman's terms, his warning suggests that the God particle could create a black hole. And Hawking isn't the only one who thinks it's possible. Many other world renowned pysicsts and scientists have expressed similar concerns.

So, how does CERN relate to the Mandela Effect?

Many believe the experiments over at CERN are causing quantum events which interrupt and alter things in our reality. Others believe the experiments force influence from another reality or a parallel universe, ultimately creating the Mandela Effects and our alternate memories by changing this in our past reality.

This is entirely possible as we have very little knowledge and control over space-time, matter, energy, and dimensions beyond the three-dimensional reality we live in.

Humans are starting to experiment in the unknown,

but unlike throughout history, our technology is extremely advanced and it actually at the point where some of the physics they're dabbling with could have extremely dangerous ramifications for humanity.

I'm not on the "armageddon / end of the world" wagon, but there's definitely some potential danger when scientists are experimenting with the God particle in such a rushed and impatient world. I fully support innovation, but there has to be slower progress because if we ever reach the point where something goes wrong, that is the point it will be too late so safety protocol should be a priority above any other.

People have also theorized that dimensions could be merging together. That we're becoming a more civilized species and tapping into a higher dimension. Though this is less likely, it's still technically possible.

### *Parallel Universe*

The theory of our universe not being unique allows for the idea that the Mandela Effect is caused by our timeline crossing over with that from a parallel universe.

A universe mirroring our own, where events play out differently based on the decisions we make and the actions we take. Though if this is true, there may be infinite universes to account for each and everyone's individual different decisions.

Imagine a mirror image of our universe where the decisions and actions not taken in our reality are taken in a mirroring reality. Or vice versa. This would cause an alternate timeline with alternate outcomes, an infinite number of alternate timelines and outcomes.

Right now we don't know, but there's definitely a lot of substance to the discussion of multiverses and parallel universes. And not just from Mandela Effect researchers, but from some of the leading minds of our world.

Unfortunately, our civilization isn't advanced enough to find the answers yet, but maybe one day we will. Or maybe someone will help us find the answers faster.

With current technology, we can only observe our universe and not physically access it or see into any other universe (if there is more than just ours).

There could be many parallel universes with alternate timelines that play out differently to our own and there are many ways the timelines could be crossing into ours to change things in history, whether recent or distant.

If that's the case, the Mandela Effect may just be the beginning.

Whether scientists are affecting our own timeline by experimenting with particles and energy we don't fully understand or whether we're unknowingly causing things to make a disturbance in space-time or another dimension is still unknown.

The parallel universe and multiverse theory is catching many people's attention and the great minds behind the idea of it being real are giving the theory more credibility.

The changes in our reality, if that is in fact what's happening, could be caused by many things from merging dimensions, space-time, gravity, quantum events, black holes and even energy. Any of which could theoretically force a parallel universe's timeline to merge with ours, for a moment to make a brief change in our history.

## *Artificial Intelligence and Simulated Reality*

Some of the biggest questions in science that surround a potential simulated reality and artificial intelligence still leave scientists speechless with no credible answers. The reason being is that they're very possible, even probable, but there's still not conclusive explanation to them despite many valiant efforts.

With the exponential growth of artificial intelligent and it seamlessly making its way into our lives through the use of technology, self-driving cars, apps, smart phones, etc, many people are getting more and more impatient in the pursuit to the answers of the most difficult questions;

Are we living in a simulation?
How would we know?
How would we escape?
Are we being controlled by a more advanced civilization?

When *The Matrix* was released, I would have laughed at these questions in a serious conversation, but now I know they're as possible as what we've already achieved in AI. I'll try and bridge the gaps of understanding the impossible for those who haven't been following the cutting edge of artificial intelligence.

Many believe we're living in a simulation, but for those who either don't think it's possible or think it's completely absurd, you might think differently shortly.

Let's just assume for a moment that we're living in a simulated reality. What does that even mean? Are we players in a video game with a super-intelligent race holding a handset and controlling us as characters without our knowledge?

That's extremely, extremely unlikely, but I'll use that analogy to explain it in easy-to-understand terms.

When we play a video game, our game can progress without us actually doing anything or even playing. For example; Tamagotchi, HayDay or World of Warcraft. Such games execute instructions while you're not even actively playing. You instruct your character(s) on where to go and what to do, and although they may carry out the tasks, they're capable of doing their own thing, at least to a certain extent. They act and react based upon the data they've been programmed with and the data they've collected. That's basic programming that allows the game's characters the ability to improvise and act based on the data they've acquired.

I'm sure you've gone back to a game at some point and realized something's happened that you didn't do, for example, with *World of Warcraft,* your characters may have been attacked and they defended themselves and anticipated what the best "move" was based on what they "know" and how you previously played.

This simply explains that a character in a game can act on their own accord to a certain extent and though the characters aren't conscious like humans, they assume they're operating under free will within the confines of the game's rules.

Now let's fast forward to a brand new game where we're getting extremely close to photo-realistic graphics. Assuming any rate of improvement, even centuries into the future, thought it'll likely be in the next few decades, humans would have created a simulation that was so flawless it's completely indistinguishable from our own reality. This isn't too difficult to imagine as we're nearly at that point already. We'd have characters we created in a computer simulation that have emotions, thoughts, organs, consciousness and they would have absolutely no idea they're in a simulation we created.

Then comes virtual reality which allows us humans to "step into" the simulation we've created and technically become a part of it, similar to the novel and movie, *Ready Player One*.

Now, who's to say someone didn't beat us to it and that hasn't already happened? There's actually no way for us to know. What we consider free will could be us operating within the rules of our simulation. Gravity is a good example. We're unable to defy gravity within the rules of our physics as we understand them, but if an advanced civilization created the simulation we live in, they may be able to totally defy the logic and understanding we live by, such as fly without propulsion, similar to the theoretical extra-terrestrial flying saucers that can achieve silent take off and move to impossible speeds without chemical propulsion.

So what's the chances we're in a simulation? That's the million dollar question that I actually have the answer to. The chances of us living in a simulation is an exact 50/50 and I'll explain why...

If we're living in a simulation, then the civilization that created our reality may also be living in a simulation that was created by an even more advanced civilization. We could be the third simulation down the line or the billionth.

Either we're base reality and we're not living in a

simulation at all or we're the civilization in the chain that has yet to create photo-realistic simulations. There's no other option which leads to the conclusions it's a coin flip whether or not we're living in a simulation. If anyone has a better argument, please feel free to e-mail me your thoughts as I would love to hear them (my e-mail is at the end of the book).

With this is mind, the next question that may arise is "how could we escape our simulated reality?"

Well, one way would be to consider how an intelligence in a simulation we created would access our reality. I can only think of one feasible route to escaping a simulated reality made by a more advanced civilization. Like humans who have been developing AI for years, that AI started with simple intelligence such as calculator, but now it's starting to approach ASI level.

Once an artificial superintelligence is created in a simulation on our computers, it would not only be more intelligent that the smartest human, but it would be able to become more intelligent than all of humanity combined. From creation to becoming smarter than all of humanity combined could happen as quickly as a few days. This worries some of the smartest minds working on the cutting edge of AI.

If you'd like to learn more about AI and an artificial super-intelligence, I'd suggest reading a book by Nick Bostrom called "Superintelligence" which you can find here: https://amzn.to/3944Nv4

An artificial superintelligence could penetrate our reality from the digital world by using the internet and computers to break free and take control of our world, for better or worse.

It would be able to flood itself into every cell phone, computer, vehicle and device on earth in a matter of seconds. This could be to help improve the systems and sling-shot our technology decades into the future or it could be for a more dystopian reason.

Using this logic, we could try and merge with a digital superintelligence and potentially extend our consciousness beyond the human body. In brief, this could be done by duplicating our neural network and basically living forever as a machine without the physical constraints of a biological body. For more on this, you can look at Elon Musk's Neuralink (https://neuralink.com).

This would allow us to process, learn and develop at unimaginable speeds. We'd be super-intelligent cyborgs who would look at $20^{th}$ century human intelligence as pathetic. This could be a route to escaping our simulation if we are in fact in one. If we're not living in a simulation and we are in fact base-reality, we could discover the way to prove that.

Going back to the Mandela Effect... If you can wrap your head around the idea of us living in a simulation, the thought of minor changes in our reality one of the easier things to imagine.

Just like us changing a logo or adjusting the

spelling of a word in a simulation we've created, that's what an advanced civilization could be doing in our reality to cause the Mandela Effect, if we are in fact living in the Matrix.

To them, it could be the simplest of things, editing a file, deleting a file, maybe even making a change to our reality with a mere thought, but to us it's a worldwide phenomenon that puzzles millions of human beings.

# Chapter IV:

# The Mind, Misdirection and Illusion

Many people whom have done a little deeper research into the phenomenon seem to attribute the Mandela Effect to cognitive dissonance.

Cognitive dissonance is the scientific term for when a person holds two or more contradictory beliefs at the same time or when a person is confronted by new information that conflicts with their existing information.

It's true that our brains usually interpret things in the simplest way so we can process them faster and increase our input of information, this can sometimes lead to us processing information incorrectly or overlooking information.

This would be a great explanation to the Mandela Effect as they're technically false memories that could be caused by cognitive dissonance, especially if the person was under stress or mental discomfort at the time.

Unfortunately, the Mandela Effect examples are different from false memories because large groups of people experience the same connection.

Another theory around the Mandela Effect being a memory issue is confabulation, a term which is more commonly used in the psychiatric field to describe a disturbance of memory, defined as the production of fabricated, distorted or misinterpreted memories.

People who confabulate recall incorrect memories and are very confident about their recollections, despite contradictory evidence being presented to them to show they're misremembering.

Again, this would be a great explanation for the cause of the Mandela Effect and alternate memories. There's just one problem with both of these theories...

Even though millions of people around the world could be experiencing cognitive dissonance or even confabulation at the same time or over periods of time, why do all these people have the same memories relating to the exact same events that apparently never happened?

If each person had unique alternate memories and a couple of people agreed with them from time to time, maybe the Mandela Effect could be the outcome of said conditions. But when millions of people are having the same alternate memories of the exact same thing and in some cases everyone's remembering a Mandela Effect example in the exact same way from what it is now, then it seems nearly impossible to comprehend that either is the reason for the Mandela Effect.

So, what's the real reason for the masses of people who are having the exact same alternate memories?

Cognitive dissonance and confabulation are likely a small contributing factor, but most of the blame has to fall on the human brain.

Let me start to explain this with going over a few categories.

## Misdirection

A magician's best friend is misdirection and this provides a clear explanation of how easy it is to trick the human brain. In short, misdirection is having a person or group focus on one thing to conceal the truth in order to reveal something they didn't expect.

Misdirection can be done on purpose, for example by a magician. It can also happen without anyone trying to misdirect you. You'll see some very good examples of this in the alternate memory explanations chapter.

Harry Houdini, the great illusionist and escape artists once shared a phrase that went something like this: "What the ears hear and the eyes see, the mind believes".

Houdini understood that the brain could be tricked by showing people's eyes and filling their ears with something other than what was actually happening, allowing him to completely confused his audience with illusions that were achieved by misdirection.

Houdini would be shackles underside down in a sealed tank that was filled with water. Within 100 seconds he did the impossible and escaped. And let's not forget his straight jacket attempts or other illusions.

The audience watched on and after he revealed the "magic" that occurred during the misdirection, everyone had the exact same reaction. They accepted and believed what they saw and had no doubt they knew what they were witnessing, but then they were speechless at the result because they had the truth concealed from them. Using this analogy we can adapt it to the Mandela Effect. The brain can quite easily overlook information when processing the data that creates memories.

Our eyes and ears can deceive us and our brain processes that inaccurate data and when we recall said memories, it's based on inaccurate or overlooked information.

## Illusion

Different from misdirection that attempts to take your attention off the truth, illusion is targeted more on tricking the brain. A good example of tricking the brain is the following diagram. Whilst looking at it, just for a few seconds, ask yourself which of the two ables is longer. The left or the right;

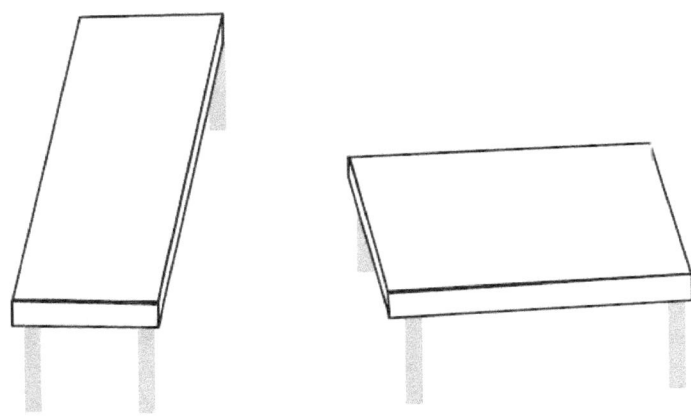

It's not a trick question and I'm sure you looked at both the tables and instantly concluded the table on the left is the longer of the two.

If that was the answer you arrived at, you're not alone. Most people give the same answer, but the truth is both tables are exactly the same length.

I'm sure you've come across the eye illusions where you spend a few moments focusing on a picture to reveal a hidden image that your eyes didn't pick up on the first time.

That's a similar concept to this that fools the brain through the eyes. As you can see in the next two illustrations, both tables are the exact same size;

With this example showing how easy it is to fool the human brain through the eyes, you can imagine how easy it could be to see something and take it at face value even though it's far from accurate or the truth.

Expanding on the magic and illusion section, I

studied magic for many years, mainly card manipulation. I would take a brand new deck of cards and ask someone to choose a card, then sign it so there was only one and it was unique.

Without any funny moves, I would fairly and visually place their signed card in the middle of the deck and instantly, I could turn over the top card and it would be their signed card. I did this over and over again in front of their eyes because what the ears here and the eyes see, the mind believes. It tricked their brain because their eyes and ears weren't seeing or hearing the truth so this blocked the brain from ever knowing the truth and it created an unexplainable magic for them.

These are just a couple of the example of how the brain can be fooled to try and illustrate the ease of misremembering something that at the time wasn't significant. And you don't need a magician or illusionist to achieve said experience because the brain can overlook the truth itself without needing any help, as with the Mandela Effect.

## Logical questions

Another area I would like to share is something called logical reasoning questions. These are questions with extremely simple and logical answers, but most people struggle to find the answers. Here are a few logical reasoning questions that will trick your brain. Remember, these are not trick questions. They have very straight forward and logical answers. I'll add the answers at the end of the chapter.

1. How far can a dog run into the forest?
2. A red house is made of red bricks. A blue house is made of blue bricks. A green house is made of ..... ?
3. You dig a hole 3 feet x 3 feet x 4 feet. How much soil is in the hole?

If you answered all three of those immediately and correctly, you're in the 0.01%. I used these examples as the masses will not be able to answer all three questions as well as the table illusion, but this was more for illustration purposes to give a collection of examples of how easy it is to trick the human mind.

## Processing

Processing information, which I touched on earlier, is one of the main reasons I feel people are misremembering actual events. It's said that the average person sees 2,000 ads per day. And in such a rapidly developing world where there's more data to consume, our brains must work much faster than they would have needed to a century ago. Especially with how easy it is to consume more data with the technology and access to the world in our cell phones.

As you can imagine with more to process, accuracy and efficiency may be sacrificed to keep up with the mountain of data coming through the ears and eyes each day.

Our brains are a muscle and the framework is the same as almost any other human's. This could account for some of the alternate memories that people are having on a large scale as we would all process such information similarly. Remember, with a lot of the alternate memories not everyone misremembers them. It can range from a thousand people having the same alternate memory to almost everyone.

This seems to also depend on the complexity of the alternate memory, but you'll learn more about that in the explanation of each alternate memory.

## Poor memory

By now you're probably already losing confidence in how accurate and efficient your brain and memories really are.

Many of us haven't got perfect memory and even if we have, we're still susceptible to forgetting details. Let's take a random alternate memory. *Snow White and the Seven Dwarfs*. You're adamant the Evil Queen spoke into the mirror and said the line "Mirror, mirror on the wall", right? Cast your mind back to the last time you actually watched the movie. For me, it was a long time before I learned of this alternate memory people thought they were having.

We'll look at this example a lot more in the explanations, but recalling a single word being slightly different from a movie you haven't seen in potentially decades doesn't mean there's been some magical change in our reality. It's simply the brain recalling data incorrectly or inefficiently. Whether we like it or not, that's the human brain.

How many times have you lost your keys? Have you ever gone into a room or a store and forgot why you went in there from literally seconds ago? Have you ever said "I've done it again" making the exact same mistake twice?

How can we possibly forget something from minutes or hours ago like where we set our keys down, but we can blame recalling a word from years ago incorrectly on something other-worldly?

Logical reasoning answers:

1. A dog can only run half way into the forest, then he'd be running out the forest.
2. A green house is made of glass.
3. There is no soil in a hole.

# Chapter V:

# Accepting the Impossible

The truth is, the Mandela Effect comes down to a simple case of memory inaccuracy. Even though we've looked at why and how that happens, many of us still want to believe it's something magical that can't be explained because like learning the secret to a good magic trick, the suspense and excitement is over.

Forrest Fenn, a wealthy art collector from New Mexico knew of this concept well and put it into affect for many adventurers.

Fenn hid millions of dollars of treasure somewhere in the Rocky Mountains in 2010 and gave clues to its location in his book titled *The Thrill of the Chase* (https://amzn.to/395v4t6).

It was a mystery for years whether he was telling the truth and if he was, where he hid the treasure. I actually spoke with Fenn personally and the thing he enjoyed the most was the adventure of families and friends prospecting for the mysterious treasure.

Fenn's goal was to provide adventure to others in

hopes some lucky person would find the treasure before he died. Many called it a hoax. Many searched. One person even attempted to file a law suit against him for misleading them.

Through everything, the truth was revealed. In June 2020, the clues were solved and the treasure was found. Fenn completed his mission of injecting excitement and adventure into people's lives and three months after the treasure was found, he passed away.

Though many of us want to know the truth and conclude the baffling mystery of the Mandela Effect, there are many of us who also enjoy the thrill of the chase and it's those who prefer to not have the logical answer because it's far less exciting than chasing the rainbow of mystery.

But, now we've discovered the causes behind the Mandela Effect, whether you're satisfied with the answer or not, it's up to you whether you want to prove the theory wrong by finding another-worldly explanation.

Combining the ease of tricking the brain with illusion, how we can be so easily misdirected and how we can forget where we put our keys from five minutes ago with the issues, conditions and how fast the brain has to process and recall data, it seems fair that by swallowing our pride we can conceive to the Mandela Effect being caused to brain inaccuracy and inefficiency.

Finally, the part of the brain that recalls these

memories we so strongly believe are true is the same part of the brain that also recalls how the memory made us feel, as well as the sights, sounds and smells associated with the memory. This part of the brain is called the hippocampus.

It's the same part of the brain that also reconstructs the past as well as allows us to imagine ourselves in the future, suggesting that the part of brain that recalls these all-so-accurate memories can build the past and imagine.

I feel all this would be a fair conclusion as to why we're experiencing the alternate memories around the world, but to put the icing on the cake and push people over the edge of belief that it's solely our brain misremembering, we'll look at some of the top Mandela Effect examples in detail to give a little more peace of mind that we're not having our reality tampered with by another dimension or having our timeline cross with another timeline from a parallel universe.

# Chapter VI:

# Alternate Memories Explained

Now we're equipped with the knowledge that are brains are not quite as accurate as we first assumed, let's clear the final part of the Mandela Effect up once and for all and explain each of the alternate memories in detail.

### Nelson Mandela

The number one alternate memory to handle has to be the death of Nelson Mandela. When Mr. Mandela died in 2013, it shocked everyone who previously believed he'd died around 1990, but why did people even think he died back then?

In addition to the book verse mentioned earlier, there was no widespread internet to fact check or correct rumors, an ability we take for granted today. With a few misinformed people and some word of mouth, it's easy to see why many people built false memories of his death. Unless you followed South African politics or Nelson Mandela, which I'm guessing most people reading this didn't, then it would be easy to miss the fact he was released from prison and became president.

Another reason would be that he was battling for his life in hospital in the 1980's and his wife spoke publicly about the unlikeliness he would survive, but he did.

With people remembering a televised funeral and rioting in the streets after his death, this is also quite easily explained.

There were large numbers of riots in South Africa around the time of his false death, here are some from just 1990-1991 to illustrate just how many incidents people may be recalling, and misremembering them as the aftermath of Mandela's death.

### 1990

22 July, 19 people are killed, allegedly by Inkatha Freedom Party (IFP) in collusion with South African police during the IFP launch at the Sebokeng Stadium in Vaal.

23-25 July, In retaliation of the killings of the 22 July in Sebokeng, 30 people are killed in the area allegedly by Inkatha Freedom Party (IFP) members.

1-11 August, 13 are killed in Sebokeng, allegedly by IFP members in collusion with South African police.

5-23 August, About 122 Soweto residents are killed, allegedly by IFP members in collusion with South African police.

12-15 August, About 150 Tokoza residents are killed, allegedly by IFP members in collusion with South African police after forced expulsion of non-IFP from hostels and attacks on Phola Park.

14 August, 24 Katlehong residents are killed, allegedly by IFP members in collusion with South African police during pre-dawn attacks on Crossroads squatter camp.

1-2 September, 44 residents in Tokoza, Tembisa and Vosloorus are killed during attacks on townships.

4 September, 19 hostel dwellers are killed, allegedly by IFP members in collusion with South African police.

4 September, 11 hostel residents are killed allegedly by IFP members in collusion with South African Defense Force.

8-9 September, 26 residents in Tladi township are killed allegedly by vigilantes.

28 October, 16 residents of Naledi Township in Soweto are killed, allegedly by IFP members in a revenge for the IFP member who was killed.

15-19 November, 34 residents of Katlehong, a township also located on the outskirts if Johannesburg are killed, allegedly by IFP supporters in an attempt to take over Zonkizizwe squatter camp.

26 November, 13 Dobsonville hostel residents are

killed allegedly by vigilantes.

26 November, 11 Katlehong residents are killed by vigilantes during a night attack at the Mandela View squatter camp.

2 December, 30 people are massacred in Tokoza a township on the outskirts of Johannesburg, allegedly by IFP supporters as political violence intensifies.

3-8 December, In an ongoing series of attacks and revenge attacks 33 people are killed in Tokoza, East Rand as political violence escalates.

11 December, 52 people are massacred in Tokoza Township, allegedly by IFP supporters.

**1991**

12 January, 45 African National Congress (ANC) mourners are killed in Sebokeng, allegedly by IFP members in collusion with South African police while attending a funeral vigil for ANC member.

3 March, 24 IFP members are killed, allegedly by Xhosa speakers in Meadowlands, Soweto in an attempt to take over the Mzimhlope hostel.

24 March, 12 African National Congress (ANC) supporters are killed in Daveyton, Gauteng, allegedly by South African police. The incident unfolded when police was dispersing an 'illegal' gathering.

14 April, 11 Nancefield residents are killed following a Clash between Nancefield hostel and Power Park squatter camp.

28 April, 10 IFP mourners are killed in

Meadowlands after a service for assassinated IFF leader.

23 May, 13 Sebokeng residents are killed while in a beer-hall.

8 September, 13 IFP members are killed under mysterious circumstances in Mofolo, Soweto.

8 September, 23 IFP supporters are killed in Tokoza by 3 unknown gunmen as political violence escalates.

13 October, 10 Mapetla township residents are killed by unknown gunmen in a tavern.

Concluding that Nelson Mandela didn't die in the 80s or 90s, and explaining why people are misremembering his false death, it shakes the entire Mandela Effect phenomenon to its core, but we're far from finished and I'm sure many of you are desperate to have an answer to each of the most mind-boggling alternate memories millions of us have experienced, so here we go...

## Empire Strikes Back

The epic like delivered by Darth Vader in Empire Strikes Back has been a hot topic for Mandela Effect researchers and Star Wars fans as well.

Most people recall Vader saying "Luke, I am your father", but the line actually goes *"No, I am your father"*.

Let's start by analyzing a little more of the screenplay to get some context on why the line has always been the way it is today...

Vader: *Obi-Wan never told you what happened to your father...*

Luke: *He told me enough. He told me you killed him.*

Vader: *No. I am your father.*

It's not too difficult to argue that Vader responding with "no, I am your father" is a more natural response because he was correcting Luke's assumption by saying "no", then announcing that Vader is actually his father. In that conversation between the two of them, Vader replying by saying "Luke, I am your father" just doesn't make as much sense.

Shortly after this scene, Vader uses the force to communicate with his son and says *"Luke"*, to which Luke responds *"Father"*.

Many people may be recalling these two combined memories and assuming the line was "Luke, I am your father".

At that point of the movie when the line was said was one of the biggest moments in cinematic history. The focus was more so on the big reveal and twist than it was on whether Vader said "Luke" or "no".

With a few people misremembering the line when the movie came out and reciting the incorrect line, it would have stuck better than the actual line and spread like wild fire with everyone saying "Luke, I am your father" as apposed to the actual line.

The difference in the two lines is minimal, so overlooking the accuracy to the movie when it sounds good wasn't even a consideration.

Without an online community of people to reach out to and agree on the accuracy of the line, it's possible many people just accepted the false line as accurate to the movie and over the years it's naturally embedded itself in our culture.

Lastly, the original Empire Strikes Back book that was released around the same time has the line delivered to Luke as: *"No, Luke. I am your father."*

## Chick-Fil-A

Considered one of the most confusing alternate memories of the entire Mandela Effect phenomenon, at least by those who're familiar with Chick-fil-A, the true name of the fast-food company has left millions speechless.

This alternate memory was one that I simply couldn't believe and it forced me to research deep into the heart of the Chic/k-Fil-A battle, even to the point I contacted Chick-fil-A.

There are many people who remember it as Chic-Fil-A, many who remember it as Chik-Fil-A and others that remember it correctly, as Chick-fil-A. This already weakens the foundations of this alternate memory.

When this false memory started to surface online, Chick-fil-A actually purchased "chicfila.com" and redirected the domain to their official website "chickfila.com" to capitalize on the massive search traffic.

Furthermore, the calligraphy font of the logo makes it very easy to process as "Chic-Fil-A" and ignore the "k" in the title. And then of course there's the printing of the millions and millions of Chick-Fil-A boxes that cuts the "k" off;

Is it possible that our eyes have once again deceived us? Is it possible that the "Chic" references and jokes over the years have just made us subconsciously believe the name was "Chic-fil-A"?

Chick-fil-A was registered in 1963 and the first location was opened in 1967, the name was registered at "Chick-Fil-A" and has been the same ever since.

Google Trends allowed me to search as far back as 2002 and there were no searches for "chic-fil-a", making it seem as if no one searched the incorrect term until the Mandela Effect raised awareness of the Chic vs Chick battle.

## Berenstain Bears

With just a simple letter being the cause for alarm, I don't feel this false memory warrants too much attention. One of the Berenstain family members said *"Of course it's always been Berenstain"*.

This alternate memory is not worthy of the credit it gets. Some remember it correctly and some remember it as "Berenstein". It's just a processing issue. Some may find it easier to assume it's "Berenstein" instead of its true title "Berenstain".

Over there years, there have even been prints of the mis-spelled title which has only fueled the fire. One of the examples was on a list that has been bashed around the Mandela Effect community and on Youtube as if it's some kind of conspiracy theory.

This is laughable. Someone in the graphic design pr printing department just spelled a single letter incorrectly and it went to print. A few times.

Here's one example:

I'm sure the person responsible for the mis-spelling got an ear-full from their boss. The confusion over the false memory is simply explained by misprints, inaccurate brain processing and maybe even the font.

## Snow White and the Seven Dwarfs

Here's one that took me a little time to actually accept. It's one of the top examples and it's earned as much fame as the other top alternate memories, but like every example of this phenomenon, it's time to end the fairytale and conclude the magic mirror on the wall.

Why does nearly everyone on planet earth remember the line in *Snow White and the Seven Dwarfs* going "Mirror, mirror on the wall" when the actual line was "Magic mirror on the wall".

The animated movie was released close to a century ago and when was the last time you actually saw it?

I can almost guarantee that most people have heard the "mirror, mirror" line recited far more recently than they have seen the movie. I know that was the case for me before I started to research this memory.

This would easily influence us to believe the line is "mirror, mirror". I remember when I was younger I was watching a play when they recited the line as "mirror, mirror" instead of the actual like "magic mirror". I've heard "mirror, mirror" many times since and I just accepted that was the actual line from the movie.

It' been hardwired into our culture and brain over the years and even those who went back and re-watched the movie with the knowledge of the alternate memory would like have just glazed over what the evil queen said.

But if the line is in fact "Magic mirror" and not "mirror, mirror", as that has always been the case then where did "mirror, mirror" come from?

The original Brother's Grimm tale, Little Snow-White by Jacob and Wilhelm Grimm, tells the story of a King's wife, the Queen, who stands before her magic mirror and asks *"Mirror, mirror on the wall, who in this land is fairest of all?"*

This line has been recited in plays, theater, other movies, book, shows and people throughout the years, so you're not necessarily having an alternate memory of hearing the "mirror, mirror" line a thousand times, you're just misremembering that it was in the 1938 animated movie.

## Jaws and a Bigger Boat

Chief Brody, played by the late Roy Schieder, says a line in the movie upon realizing the shark they're trying to battle. His line in the movie is *"You're going to need a bigger boat"*.

A good portion of the Mandela Effect community remember the line as *"We're going to need a bigger boat"*, which is logical considering Brody is on the boat and they're trying to fight the shark together.

However, it's just as easy to assume he said "You're going to need a bigger boat" as he is talking directly to the owner of the boat and maybe wants to isolate himself from the shark-hunting group out of fear. Either line could work in the context he said it.

The movie is nearly half a century old and upon initial release, the sound quality wan't quite the 5.1

dolby digital we had today, so hearing the line would not only have been slightly more difficult with Brody's mumble of fear, but with a cigarette in his mouth and the ambient noise in the background, re-listening to the line could easily be perceived as either "we're" or "you're".

Not only that, but people who misremembered one word from the line have created merchandise that says "we're going to need a bigger boat" which has also helped develop a false memory of the actual line.

## Interview with the Vampire

Interview with the Vampire was one of the earlier alternate memories people reported with millions searching for the movie online as well as the novel by Anne Rice to double-check the spelling of the title.

Though just a single word has caused this explosion in alternate memory reporters, many people are adamant the title was always "Interview with a Vampire".

We should be able to sweep this under the rug by stating people just misremember "the" for "a", but that wouldn't conclude this false memory well enough, so let's go a little further.

The title of the movie in Germany is *Interview mit einem Vampir*, which directly translates to "Interview with a Vampire", making people's false memories accurate if they're basing their memory off the

German title. The French title also directly translates to "Interview with a Vampire". Why are the foriegn translations relevant to us? Well, the Mandela Effect reaches globally and France and Germany are big contributors to the movie community, both festivals and online, so it's easy to see how this could cause a snowball effect.

In 2017, even Amazon employees who listed various versions of the book and movie sometimes misspelled it "Interview with a Vampire" despite the title being "Interview with the Vampire" and the Amazon product title had to be changed to the correct spelling.

## Sex and the City

Another popular alternate memory that has people questioning their sanity is the "Sex in the City" vs "Sex and the City". Ironically, this is actually the easiest one to explain.

The perfume is titled "Sex in the City" and the hit TV show is titled "Sex and the City".

## The Monopoly Man

This is probably the explanation most people came here for because I have seen countless complaints about this alternate memory and people are just losing their minds about Uncle Pennybags not having a monocle "anymore". I know because I was one of them. This is going to be hard to hear, but Uncle Pennybags has never had a monocle.

It's the truth, but it's not easy to hear that your brain is misremembering every single recollection of a Monopoly board, money note, logo or artwork, so I'll try and take the edge off for you.

As popular as Monopoly is and as many times as we've seen it and played it, you'd think that we'd remember iconic characters perfectly right? Using the same brain that can't remember where we put our car keys five minutes ago...

Not everyone misremembers the monocle. Many remember his always being without a monocle, but what about the millions of people who do remember him with a monocle? Well, I'm guessing you've either seen the movie *Ace Ventura: When Nature Calls* or you've seen some kind of reenactment of someone with a monocle in their eye and put those two memories together.

In the 1995 Jim Carrey classic, there's a scene where an older man walks down the stairs in a black and white suit. And yes, he very accurately resembles Uncle Pennybags, even to as accurately as having a mustache and a bald head, but that's not all. He's also wearing a monocle.

Oh, and Ace Ventura says *"you must be the Monopoly guy"*, then he knocks him unconscious and shakes him to act out him saying some of the famous lines from the board, such as "Do not pass go".

Even if you don't remember watching the movie or have seen it and don't remember anything from it, it's still completely possible that our hippocampus is recalling that data when thinking about Uncle Pennybags and making us belief the actual "Monopoly guy" always had a monocle when he never has.

There are many other characters that millions of us has crossed paths with who dress exactly like Uncle Pennybags and do wear a monocle. A few examples would be the Penguin from Batman Returns, Mr. Peanut, and the Mayor from the Powerpuff girls.

Uncle Pennybags is also the kind of character who would suit having a monocle. Monopoly was first released in 1935 and Uncle Pennybag's attire was inspired from that time, naturally a monocle would have complemented his fashion perfectly and with everything else we've discussed, we see that monocle on him in our memory even if it's not there.

Monopoly heard about the phenomenon and the alternate memory of Uncle Pennybags and decided to pitch in by sharing a picture of him with a monocle on Facebook to tease those who had the false memory which you can see here: https://goo.gl/gC1yt3

## Jaws and Dolly

Here's one that is not only driving Mandela Effect followers crazy, but it's going to take some explaining because there are many contributing factors as to why millions of people are having an alternate memory and misremembering Dolly from Moonraker having braces.

In the 1979 movie, Moonraker, Jaws' love interest, Dolly, never had braces. Thousands of people are absolutely adamant she did, but why do so many people falsely remember this?

The first report of this inaccurate memory was on August 2, 1999 and is available for you to view here: https://goo.gl/TEvXy5

Over the last couple of years, the alternate memory has received a lot of attention and rightly so because it would seem her having braces was the key reason Jaws, who had metal teeth, was attracted to her.

I still find this difficult to digest because I grew up on Bond movies and saw Moonraker a good three or four time over the years, but here it is...

1. People seem remember Jaws and Dolly's bond beginning when she smiled at him. This is correct, but she was never wearing braces. If you go back and watch the VHS, which is the most popular format people are recalling their memories from, you'll see a shine over her teeth from the 450x450 pixel quality and it could be perceived as her wearing braces if you weren't paying direct attention to her teeth.
2. In the scene where she's drinking a glass of

champagne with Jaws, the image you see of her mouth through the glass really does appear as if she's wearing braces, but she isn't.
3. There have been many websites who have profile Dolly over the years and referenced her with having braces which has added to the false belief that she wore them in the movie.
4. The final reason for many people, including myself, are misremembering the truth was due to a 1991 TV commercial. Although none of the commercial would ever warrant a place in our memory, there's one thing that would. The actor who played Jaws, Richard Kiel, starred in a Visa commercial that promoted the mini-card.

The cashier who co-starred with Kier in the commercial looks strikingly similar to Dolly from Moonraker. The camera angle of the cashier's first appearance on camera was almost identical to the camera angle of Dolly' first appearance when she met Jaws in the movie. It was clear they were trying to reenact that moment from 12 years earlier. And yes, she smiled at Kier the same way Dolly did. And yes, she had braces.

Combining all these pieces of the puzzle together

with everything else we've been over, it's easy to see where the confusion has come from.

## Three Little Pigs

This is another rather simple explanation for people who are having a tough time figuring out why they're having an alternate memory of the classic children's nursery rhyme.

The debate over this alternate memory is whether the line was "I'll huff and I'll puff and I'll blow your house down" or "I'll huff and I'll puff and I'll blow your house in". Another one word alternate memory, but it's a fair request as many of us have recited the story / line hundreds of times.

In the UK, we were raised on the "I'll huff and I'll puff and I'll blow your house down" version and that's how many people remember it.

In the US, it seems that people were raised on the "I'll huff and I'll puff and I'll blow your house in" version.

After a little more research, it seems that there's different versions for different geographical locations and with there being a book, song, poem, different dialects in different regions, it's easy to see why people think they're misremembering this.

It's also possible that a version with "in" made it to a part of the world that was used to the version with "down".

# Hello, Clarice

Here's the Mandela Effect example that got me straight on Youtube and had me typing as fast as possible to search for the scene from *The Silence of the Lambs* where Hannibal Lecter greets Clarice during their first meeting by saying "Hello, Clarice".

Unfortunately for me, I was left with Hannibal Lecter's introduction to Clarice being a simple "Good morning". This hurt. The next step? Watching the movie from start to finish to figure out where in the movie he actually said that line. Again, I was kicked in the face by the Mandela Effect as Hannibal, at no point in the movie, says "Hello, Clarice".

I was absolutely adamant I heard that line on screen and I was not willing to concede. Well, that line was said on screen, but not in *The Silence of the Lambs*.

Once again, we have Jim Carrey to blame for poisoning our memories with a parody scene from The Cable Guy in 1996.

We he and his forced-friend visit the Medieval Times restaurant (which is actually a brilliant place to visit by the way), he uses the skin from his turkey leg to place over his face. This directly mimics the scene from *The Silence of the Lambs* when Doctor Lecter poses as a police officer by eating the skin from his face and placing it over his own in order to escape in an ambulance.

Yeah, you guessed it. Jim Carrey's line at the restaurant is *"Hello, Clarice"* in his best Anthony Hopkins impression.

The full line he says is actually:

*"Silence... of the Lambs... Hello, Clarice. It's good to see you again."*

## C-3PO's Leg

This alternate memory has left a lot of people looking for an explanation, especially Star Wars fans who're extremely attentive to detail and have literally watched the movies over and over.

The problem is, the silver leg blends in relatively well with the rest of the gold body of C-3PO and doesn't stand out at all unless you're looking for in.

Just goo back and re-watch the opening sequence in *Star Wars: Episode IV – A New Hope*. You'll barely notice the difference in color in certain light and angles so considering this a credible alternate memory is really craping the barrel.

## The Flintstones

Apparently there's many people who remember The Flintstones being spelled "Flinstones" without the "t".

I'm sure you feel safe assuming that I'm not bias towards debunking the Mandela Effect examples because I've shared how gobsmacked I've been at some of the alternate memories I've experienced, but this one really doesn't deserve to even be on the list.

The stone age began approximately 2.6 million years ago and lasted until around 3,300 BC when the bronze age began.

The Flintstones is set in 10,000 BC which falls perfectly in line with the stone age period.

Flint is a form of rock that was used throughout the stone age. Need I go on?

## Looney Tunes

This one is quite an easy explanation and no one can really blame themselves for falling victim to his false memory. There are far less people who are experiencing this alternate memory, but there are still a good group.

Simply put, the *Looney Tunes* show has always been *Looney Tunes*. People are having conflicting memories of the miniature spin-off of *Looney Tunes* which was actually called *Tiny Toons*.

It's easy to see why some are misremembering the title of the show as "Looney Toons".

## The Matrix

The Matrix has you!!!!!
This is probably how some of us felt when we

learned that the line "what if I told you" was never said by Morpheus in the Matrix during the pill scene.

Okay. Time to grab the bull by the horns and explain one of the most mind-blowing Mandela Effect examples that people all over the world are experiencing.

The answer is simple to explain, but different to believe. When recalling this line, there are many people who will stake their soul on the line being in the movie.

Do you clearly remember the scene in the Matrix where Morpheus says something similar to "what if I told you everything you believed was a lie"?

You're not alone. Many people remember this happening, but it never did. The line was never said by Morpheus, neither during the pill scene where he offers Neo the Red or Blue pill, nor in any other part of the movie or the sequels for that matter.

The answer to this mind-bending memory is a simple meme.

Just like Jim Carrey putting his own twist on Hannibal Lecter and creating a phrase that was never said and just like when people spread the "Luke, I am your father" line that was never actually in the movie, one person created a meme that started with "what if I told you", followed by a long list of different phrases that made it one of the most popular memes of all time and gave many people the false conception that this line was actually said in the movie.

Now we've learned about all the inefficiencies of the human brain and compiled that information with the explanation for the alternate memories mentioned earlier in the book, it's time to cover a few more alternate memories I've seen reported over the internet so anyone searching for answers isn't left stranded.

However, if I've missed any alternate memories out that you'd like to discuss and have explained or if you'd like to have any further clarification or conversation regarding the Mandela Effect, I'd be happy to speak with you via e-mail.

Let's rapid-fire more reported examples from the Mandela Effect...

### Alice in Wonderland – We're All Mad Here

The Cheshire cat in the Disney classic, *Alice in Wonderland*, is remember by a large number of people as saying "we're all mad here", some of which have even gone as far to get the phrase tattooed on them.

The actual line in the movie is *"most everyone's mad here"*. So why is everyone misremembering the line?

There are actually two variations of the line ("we're all mad here" and "we are all mad here"), both of which are twisted versions of the original line and neither are in the movie.

One important thing to note is that the 1951

animated classic is not the only version of the movie and it's not the original story either. The source material actually dates back to nearly a century before the animated classic was released.

Lewis Carroll first wrote the novel in 1865 which was titled *Alice's Adventures in Wonderland* (which you can see here: https://amzn.to/2LTwKgq)

There's at least one more version of the story in the form a book by Jane Curreth where the cheshire car does in fact say the exact line *"we're all made here"*.

With its extreme popularity over the decades, there have been hundreds of plays, versions, reenactments and parodies of the classic tale, many of which do in fact include the line *"we're all made here"*.

This has influenced millions to falsely believe that the 1951 Disney classic contained the line when that's actually not accurate.

The people who got tattoos likely did a Google search for the phrase and came to the erroneous conclusion that "we're all mad here" was actually from the 1951 animated version.

Another example is a movie titled *We're All Mad Here* which focuses on the story of the Mad Hatter from Alice in Wonderland.

Either way, the line "we're all mad here" was said by the cheshire cat in multiple versions of the Alice in Wonderland story, just not the Disney movie.

With everything we've covered so far, you can probably figure out other alternate memories that crop up using the ideas and information in this book.

Things like the KitKat logo that many believed had a hyphen between "Kit" and "Kat". The JCPenney store name people misremembered as not having the "e" in the name and many more. Some not even worth mentioning.

# **Chapter VII:**

# **Final Thoughts**

I truly hope you found the answers you were looking for and weren't too disappointed with the results that the Mandela Effect isn't the mind-bending phenomenon it's been made out to be.

The brain is a wonderful thing, but scientists say that when we recall memories, we re-construct an event from traces throughout the brain. Our memories are also adaptive, reshaping them to accommodate new information and experiences. Ultimately, we could consider our memory as somewhat flexible.

We each experience alternate memories or false memories all the time, but we just discard them as simple mistakes. It seems when lots of other people mistake the same thing it becomes somewhat of a conversation piece and that's snowballed into the Mandela Effect.

Another example is mishearing song lyrics. There's a brilliant stand-up comedian named Peter Kay who did a piece on "misheard song lyrics" that you can search on Youtube. With comedy, he explains misheard and misremembered lyrics.

There are so many things we miss and so many

things that we even do on auto-pilot that our brain doesn't retain. For example, locking the front door. How many times have you got out of bed to check if you looked the front door or turned a device off? If you have, I'm sure you've found that you got out of bed and your door was locked, but you can't for the life of you remember whether or not you did it.

A even more bizarre experience that most everyone has experienced is being in the middle of a conversation and you literally can't remember what you were saying, hence the famous phrase "I just lost my trail of thought".

There are a long list of examples that illustrate how our brain can become confused or completely forget something from seconds ago.

There are even things that we've done a thousand of times over and after some time passes, we could go back to do that activity and completely forget part or all of the process.

The conclusion is that the Mandela Effect and all the buzz words around it aren't alternate memories, they're inaccurate memories.

In order to avoid falling victim to future Mandela Effect examples and false memories, it could be a good idea to accept the brain isn't as accurate as we think and hope.

Humbling ourselves to sometimes accept we're wrong might help us to tackle new Mandela Effect examples that people may share.

Looking at things in a logical way, looking at things from a different viewpoint and accepting the alternate memory is incorrect is the first step in discovering the truth.

I know some of us already have conflicting thoughts about the Mandela Effect and many other mysteries. We want to know the answers so we can sleep at night, but we're sometimes disappointed when the chase is over and there's nothing to wonder about anymore. Just like learning the secret to a good magic trick or illusion. You can watch it over and over and it really can look like magic, but when you know the mechanics of the illusion, the magic's over.

If you're like I am and you chase the mysteries to get disappointed when you discover the truth, don't worry because there's many more mysteries beyond the Mandela Effect that may get you lost in the rabbit hole.

A few suggestions for baffling mysteries would be Hoia Baciu Forest, Hy'Brasil and general research into physics.

Because there's no conclusive answers to such mysteries there's a good chance you'll have them burned onto your mind forever with no answers, so be warned.

I hope after going through all of these examples and learning a little bit more about how easy it can be to overlook and recall historic information from our brain, that this concludes the Mandela Effect for you.

I'd like to truly thank you for picking up my book and investing your time into reading it.

I'm now solely focusing on my vampire series and this edition of Alternate is my final writings on the Mandela Effect. It has been a pleasure to research this worldwide phenomenon and bring closure to those who were searching for logical answers.

Best wishes,
Jay Wheeler